Global Warming

Global warming is one of the key issues of our time, discussed by everyone from political leaders and entertainment celebrities to office colleagues and families around the dinner table. Few people now question the scientific evidence that the world is getting warmer. Instead, the two main questions are: 1) What is causing the rise in temperature? and 2) What should be done about it?

Significance

Both the nature and the consequences of global warming have been controversial. The debate in the popular media as well as the policy community has focused on:

- whether increased average global air temperatures since the mid-20th century could be—either partially or completely—the result of inaccurate measurement.
- whether this warming is within normal climactic variations, or whether it is extraordinary and unprecedented.
- the extent to which human activity is global warming's primary cause.

Until the late 1990s, most governments argued that there was not enough evidence to prove the theory that human activity was pushing the temperature up to justify taking steps that might slow down economic growth. But natural disasters and an increasing body of scientific evidence linking the burning of carbon-based fuels with rising temperatures has made this the dominant and growing view.

Something will need to be done—but what, and, more importantly, who should do it?

Alternatively, a smaller group of researchers believe that the cause is primarily related to a history of rising temperatures in 100-year cycles, rather than human influence. This view asserts that the earth will naturally right itself without dramatic changes on the part of humanity.

If the scientists in the increasingly dominant camp are right, politicians will soon face some hard decisions. The effects of global warming on societies and individuals will become increasingly visible and shocking, just as the idea that modern lifestyles are contributing to the situation becomes an uncomfortable reality. Something will need to be done—but what, and, more importantly, who should do it? Whose activities should be limited by new regulations or taxes? Can the international community tackle the problem of global warming together, or is it every nation for itself?

Context

Politics has been described as being all about who gets what, and when and how they will get it. In that sense, global warming is a very political issue, as only governments can decide how to share the burden of responding to it. One of the first big questions is whether rapidly industrializing countries such as China and India, whose energy usage is rising sharply, should have to make the same cutbacks as developed countries, whose use of fossil fuels over the past century helped create the problem we are now facing.

Did You Know?

- *Over the course of the 20th century, the global average surface temperature rose by 1.1° F.*
- *Snow cover worldwide has dropped by about 10% since the 1960s.*
- *Precipitation in the Northern Hemisphere went up by between 0.5 and 1% in each and every decade of the 20th century.*
- *Model simulations reveal that future warming will be most acute in North America and Central Asia (40% more than the global average).*

Source: www.twnside.org.sg

Talks about these issues have been going on for some time. In 2008 and 2009, the international community, working through the United Nations, is negotiating a new treaty to replace the Kyoto Protocol. This protocol was the first attempt to persuade countries to think seriously about reducing their carbon emissions. At a meeting in Bali at the end of 2007, negotiators agreed that developed countries should make specific commitments to limiting emissions, while developing countries should also have a role, which would vary according to their capabilities.

Only if global warming is seen as a problem that affects the planet as a whole is there likely to be meaningful progress in responding to it.

Crucially, the idea of specific commitments won the support of the United States, which had never fully agreed with the reduction targets set in the Kyoto Protocol. But Washington wants China and India to accept commitments rather than vague targets, and it wants to focus on technology that would reduce the impact of fossil fuels on global warming rather than simply forcing people to reduce their consumption.

The international community seems certain to put a great deal of effort into this area in the next few years, but the world faces a major problem as efforts to tackle global warming become more serious. Individual countries will continue to put their own interests over those of their neighbors. Only if global warming is seen as a problem that affects the planet as a whole is there likely to be meaningful progress in responding to it.

Nature of the Issue

The key to global warming lies in the balance of gases in the earth's atmosphere, which is almost entirely made up of nitrogen and oxygen. The atmosphere filters sunlight before it reaches the earth's surface and then allows some of that heat to escape back out into space. The result is a planet that can support the sort of temperatures in which humans can survive comfortably. Yet it turns out that the factor that regulates how much heat escapes is not the nitrogen or the oxygen but rather a small group of gases that make up about 1% of the atmosphere. Most of that 1% is actually water vapor, which doesn't do much by itself, but, when mixed with small amounts of carbon dioxide (CO_2), is very effective at trapping more heat inside the atmosphere.

And that's where humans come in. Using a number of techniques such as examining gases trapped in the ice of Antarctica, we know how much carbon dioxide has been in the atmosphere for thousands of years. Because the amounts are so small, scientists measure it not by percentage but by parts per million—one part per million, or ppm, is the same as 0.0001%. Between the end of the last ice age 20,000 years ago and the start of the 20th century, the amount of carbon dioxide in the atmosphere ranged from 260 to 290 ppm. By 2007, that figure reached 383 ppm,

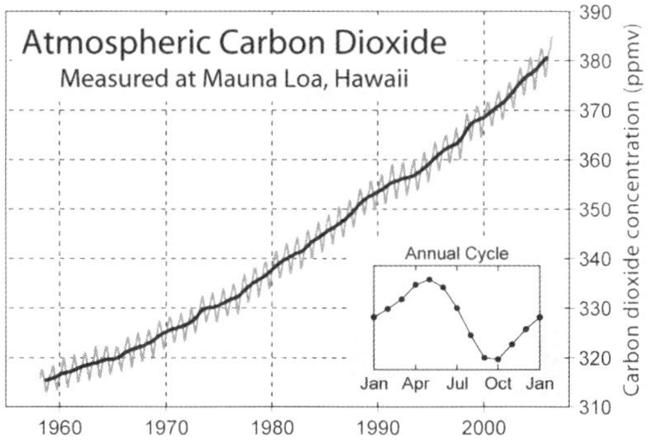

Source: www.globalwarmingart.com

indicating that an increase of about 40% had taken place over a single century.

This extra carbon dioxide came primarily from burning fuels known as hydrocarbons, or fossil fuels—oil, gas and coal. All of these have carbon as their prime ingredient, and when they are burned that carbon heads for the atmosphere as smoke and exhaust fumes. Large amounts of carbon that had been safely locked up in the ground have now been transferred to the atmosphere in, geologically speaking, the blink of an eye as the world industrialized.

Large amounts of carbon that had been safely locked up in the ground have now been transferred to the atmosphere in, geologically speaking, the blink of an eye as the world industrialized.

Even a 40% rise leaves you with a seemingly insignificant level of carbon dioxide, but it turns out that even a small rise in the level of carbon dioxide makes water vapor much more effective at trapping sunlight. As temperatures rise, more water evaporates into the atmosphere, creating more water vapor, which holds in more heat and so continues the cycle of evaporation.

Letting sunlight in and then trapping its heat is the principle behind the simple glass greenhouse used by gardeners to grow plants. These need plenty of warmth, which is what carbon dioxide, along with other gases—particularly methane, nitrous oxide and ozone—help produce. These are now known as "greenhouse gases."

Effects

This greenhouse effect has the potential to alter weather conditions such as rainfall, wind patterns and cloud cover in ways that affect global environments. That is why, as early as 1988, the U.N. Environment Program and the World Meteorological Organization set up a team of hundreds of scientists from around the world. Known as the Intergovernmental Panel on Climate Change (IPCC), its task was to collect data, create computer models of how different conditions could change the climate and provide an objective assessment of what the evidence for global warming meant. It was expected to produce a major report every six years that would provide a clear scientific context from which policymakers could develop a response to global warming.

In a series of reports since 1991, the IPCC concluded that the impact of climate change related to global warming is likely to be substantial.

In a series of reports since 1991, the IPCC concluded that the impact of climate change related to global warming is likely to be substantial. Drought will occur in more areas, affecting food production, while heavy rainfalls in other parts of the world will increase the risk of flooding. Changes to local climates will have an impact on living conditions, making it more difficult for crops, trees and fish to survive.

The first people to feel these effects will be those living in low-lying coastal areas, which will experience severe storms from unsettled weather and then flooding from rising sea levels. By 2080, millions are expected to face flooding every year due to sea-level rise, especially in parts of Asia and Africa. Millions more will face malnutrition as crops begin to fail and food supplies are disrupted, and infectious diseases will spread as clean water becomes scarcer. Potentially more devastating would be melting at the polar ice caps. If temperatures rise, the huge sheets of ice covering Greenland and the Antarctic will melt more rapidly, adding more fresh water to the oceans, which in turn raises sea levels.

In 2007, the IPCC produced a major series of reports that included both data and projections. Observations showed that, compared with the situation in 1970, spring is coming earlier in many parts of the world; water temperatures are rising in oceans and rivers; more people are dying from the high temperatures during European summers; and drier conditions are shortening the growing season for crops in many parts of Africa. Looking ahead, parts of the world farthest from the equator are likely to get much wetter, while many regions nearer the equator will receive 10 to 30% less rain. North America will be more

Looking ahead, parts of the world farthest from the equator are likely to get much wetter, while many regions nearer the equator will receive 10 to 30% less rain.

vulnerable to forest fires in rural areas and heat-related deaths in cities, as well as coastal flooding.

Viewpoints

Since setting up the IPCC, the United Nations has been coordinating international talks on climate change. In 1992, 154 countries signed up to attend the U.N. Framework Convention on Climate Change, an environmental treaty that became effective in 1994. Under the UNFCCC, countries voluntarily committed themselves to limiting and then reducing the emission of greenhouse gases. Initially, the burden was placed on industrialized countries, whose target was to produce less carbon dioxide in 2000 than they were at the start of the 1990s—quite a challenging goal, in fact, given how much emissions were rising year by year. Developing countries were not expected to do anything themselves, as tackling poverty and economic development obviously had higher priority and their overall contribution to the problem was much smaller. But some established countries pledged to provide funds and technology to begin tackling carbon emissions in developing countries.

The original U.N. climate change treaty did not set compulsory limits on greenhouse gas emissions for individual countries. Instead, it included the possibility of updates to the Framework Convention (known in diplomatic language as "protocols"), which would set compulsory limits on emissions. The first of these was agreed to in the Japanese city of Kyoto in 1997, though it took seven years to come into force. Under the Kyoto Protocol, a group of leading developed countries agreed to cut greenhouse gas emissions by specific amounts, so that the amount they emit in 2012 will be around 5% less than it was in 1990. More than 100 developing countries also agreed to monitor and report on their emissions, although they were not required to commit to any reductions.

That was a sore point for two of the leading developed countries. Al Gore, then the vice president, signed the Kyoto Protocol on behalf of the United States. But opponents in the Senate, whose approval was necessary for the United States to participate in the process, argued that any international agreement that did not require China, the world's second-largest producer of carbon dioxide after the United States, to reduce its emissions would put American manufacturers at a disadvantage. The Senate never voted on Kyoto Protocol targets, so the United States was never obligated to make the cuts in carbon emissions to which Gore had agreed. Australia also refused to accept the Kyoto targets for similar reasons: Not having to bear the costs of cutting emissions, companies in India and China would be more competitive, putting Australian jobs at risk.

Underlining the point that few countries will place themselves at a disadvantage for the sake of the planet in general, no European state wants to adopt a more stringent plan than its neighbors lest its economy lose its competitiveness.

The European Union worried about how the cost of meeting its ambitious target of an 8% reduction would affect its companies. Its response was to launch a trading system under which companies that use large amounts of carbon-producing fuels are allocated a particular number of permits. Each permit allows them to emit a metric ton (a little over one ton) of carbon without penalty. Companies that can avoid using all their allocated permits, by changing their production process, for example, or using cleaner fuels, are able to sell their surplus emission permits to companies that are going to produce more carbon than their permits allowed. The idea is to provide a financial incentive for making cuts that rewards those companies making the greatest effort.

Europe's emissions trading scheme, launched in 2005, covers about 12,000 companies in areas like electricity generation, steel production and cement, all of which are heavy users of energy. Unfortunately, while it sounded good in theory, the system was flawed. Underlining the point that few countries will place themselves at a disadvantage for the sake of the planet in general, no European state wants to adopt a more stringent plan than its neighbors lest its economy lose its competitiveness. As a result, governments tended to give out too many permits, so there was little need to trade them. And because it was not clear what sort of regulations would follow Kyoto, many companies held off making the big investments in cutting emissions that would enable them to become big permit sellers.

The Kyoto Protocol will expire in 2012, but it is already apparent that its target for emissions reduction is not going to be met. Looking back, it is clear that its goals were too aggressive and too optimistic. Governments turned out to be less committed when they noticed that others were likely to miss their targets, and the absence of the United States and China gave everyone else an excuse for taking a halfhearted approach. But during those same 10 years, the news from the IPCC became increasingly worrisome as evidence grew that the pace of global warming was quicker than it had looked only a few years earlier.

That put particular pressure on the U.N. climate control talks held in Bali at the end of 2007 to devise a successor to the Kyoto Protocol. The Bali meeting set out an ambitious schedule of talks during 2008 to consider all the issues involved, but, by general consensus, there was really only one that mattered. By 2008, virtually all parties agreed that the central problem with Kyoto was that the world's two greatest producers of greenhouse gases, the United States and China, remained outside the system. If the next protocol to the U.N. Framework Convention was going to entrench the principle of specific commitments to lower emissions, both would have to participate. It need not be in quite the same way—in a much-used phrase, these commitments would be "common but differentiated," so all countries shared responsibility for curbing global warming but would carry different parts of the burden. Washington, however, made it

Albert Arnold Gore Jr. (1948–)

The son of a U.S. senator, Al Gore Jr. spent more than 30 years in politics. He started out as a congressman from Tennessee (1977–85), then represented the state in the Senate (1985-93) before becoming the 45th vice president (1993-2001). After losing the presidential race in 2000, he focused on his long-term cause of environmental protection. His tour of the country giving a simple lecture on global warning became the Oscar-winning documentary film "An Inconvenient Truth." Gore's efforts to raise awareness of the issue won him the Nobel Peace Prize in 2007, which he shared with the IPCC.

clear that it would join in only if China and India made similarly binding commitments. This was a tough condition that seemed to go beyond party lines in America.

The report from the IPCC that appeared in three installments during the winter of 2007-08 was a powerful document. In 2007, the IPCC had won the Nobel Peace Prize for its work on developing the science of climate change. It shared the award with Al Gore, who had lost his bid for the presidency in 2000 and taken up the cause of raising public awareness of global climate change. Gore's filmed lecture on the threat of this issue had also just won an Oscar. But it was the science as much as the awards that pushed the issue up the political agenda.

The IPCC argued that it was still possible to limit the damage, and that the cost to the global economy of pursuing the necessary measures now looked much smaller than originally feared. But—and this was crucial—the world had to act quickly.

Following up on its 2001 findings, the IPCC in 2007 showed greater scientific confidence that warming in the 20th century was due to human interference and that future climate change was likely to have deeply negative effects. The report drew links between climate and society, indicating how changes in one can disrupt the other. It also presented evidence that the effect of regional climate changes can spread to all continents and most ocean areas, and that the link between global warming and human activity is particularly clear since 1970. Finally, the report examined options for limiting further emissions of greenhouse gases and strategies for reducing the impact of global warming. The news here was surprisingly upbeat, given what had come in earlier reports. The IPCC argued that it was still possible to limit the damage, and that the cost to the global economy of pursuing the necessary measures now looked much smaller than originally feared. But—and this was crucial—the world had to act quickly.

The urgency made the attitude of the United States, China and India all the more important. The ability of the world to act decisively depended on the three countries' participation in the post-Kyoto effort to tackle climate change. Once Australia signed on to Kyoto at the start of 2008, the United States was the last major developed country to remain largely outside the UNFCCC.

Several factors suggested that a reconsideration of climate policy might be possible. As is often the case, some of the states were developing ideas far ahead of the federal government. In particular, California committed itself to implementing emissions trading along the lines pioneered in Europe, while 10 major Northeast states pushed forward with plans for a carbon trading system covering electric utilities. Meanwhile, in 2007, Congress began to consider new measures that would address climate change issues. These included better fuel efficiency standards for vehicles, which cut down on a major source of emissions, and the possibility of moving emissions trading from a regional to a national level.

With global warming and energy issues increasingly dominating national news, the issue is now on the minds of both voters and candidates at every level. This suggests that the U.S. government is, or will soon be, looking for a way back into the middle of the action on climate change. But the precondition that China also commit itself to cutting emissions still has a lot of support across the political spectrum.

Fortunately, China decided to rethink its position on this issue. In 2007, its leaders finally acknowledged that environmental problems could seriously threaten long-term economic growth. The government developed a national policy on climate change focusing on reducing emissions through low-carbon energy supplies, energy-saving initiatives and agricultural changes. Although it emphasized that it would not commit to reduction targets that threatened its efforts to eliminate poverty or sustain development, Beijing acknowledged that it had a responsibility to respond to the challenge of climate change.

... [the Indian government] argued that because global warming has been caused by the actions of the developed world, it is the developed world that has the responsibility for sorting it out.

Because China's environmental protection agency is relatively weak and not able to force every province to take environmental regulations seriously, the government began to work through the much more powerful National Development and Reform Commission. In an effort to reduce the use of carbon-producing fossil fuels, China's central government gave this agency about $265 billion to invest in renewable energy sources by 2020.

By contrast, the Indian government indicated that it may not be interested in committing to a post-Kyoto treaty. It argued that because global warming has been caused by the actions of the developed world, it is the developed world that has the responsibility for sorting it out. India has one-third of the world's poorest people, whose only hope lies in pushing up levels of development: Why, Prime Minister Manmohan Singh went on, should they suffer because other countries enjoyed industrial revolution centuries ago?

Finding alternatives to the three main carbon-based fuels—oil, coal and natural gas—would probably be on the world's agenda by now even without global warming.

But, in fact, there was considerable support for environmental issues from within the Indian government—not least because the consequences of climate change would affect the nation severely. India has a strong tradition of environmental protectionism and already has some of the most stringent environmental protection legislation in the world. It stands to gain considerably from programs that move money from developed economies into developing ones in order to reduce greenhouse gas emissions. Moreover, lacking oil and natural gas resources, the governments at both the federal and state levels continue to emphasize the importance of renewable energy sources. For all these reasons, India is unlikely to abstain from future protocols.

Solutions

Politics is only part of the story. If there is to be a reduction in the use of carbon-producing fuel, that means we either simply use less fuel—by making fewer products, taking fewer journeys and using less electricity—or that we find alternative forms of energy that produce lower carbon emissions.

Finding alternatives to the three main carbon-based fuels—oil, coal and natural gas—would probably be on the world's agenda by now even without global warming. With demand for energy shooting up in China and India, pressure on supplies has pushed up prices to the point where interest in alternative fuels has increased

Reducing Carbon Fuel Use

Since the 1950s, crude oil has been the world's most important source of energy. It powers cars and aircraft, heats homes and generates much of the electricity we use. If the world, however, continues to use oil at current rates, known deposits of oil will be largely depleted by 2040. That would create significant challenges and ramifications, unless we have discovered new supplies by then; have figured out how to use oil more efficiently so that we need less of it; or have developed alternative sources of energy. The last two, at least, would also make a major contribution toward combating global warming.

Coal is made up mostly of carbon and sulfur. It is the type of fuel that the world uses most—about three-quarters of all coal burned is used to generate electricity—and this also makes it the principal source of carbon dioxide going into the atmosphere. Because coal is cheap, it is used abundantly throughout the world, especially in developing countries. At current production levels, coal reserves will last just until about halfway through the next century

Natural gas produces only half the carbon emissions of either oil or coal, which is why it is often described as clean. In addition to this, there are still plentiful supplies of it in the ground. But, being a gas rather than a liquid or a solid, it takes up a lot of room and so is difficult to store or transport across long distances. Pipelines can move it from one place to another, but they are expensive and are not practical for getting it across oceans. It is possible to shrink the amount of space natural gas takes up by converting it into liquefied natural gas (LNG), which can then be carried on specialized ships able to handle the very low temperatures involved. That, however, is expensive and requires considerable investment. Another complication is that only about 15 countries account for more than 80% of global natural gas output, which makes control of pipelines a tricky issue in international relations.

Over 85% of global energy production comes from burning oil, coal and gas, and in each case, there is a good reason to look at developing alternatives. The fact that the three fuels produce more than 24 billion tons of carbon each year gives urgency to these efforts. The European Union, for example, aims to meet 20% of its energy demand from alternative fuels by 2020. Work on several of these alternatives, sometimes called renewable energy, already looks promising.

Wind

Wind can be converted into electricity using a modern version of the old windmill design. Many parts of the world now have at least a few tall towers equipped with long blades that turn in the wind and generate electric current. Wind is clean, plentiful and pretty much everywhere, which makes it a very attractive source of energy. It already provides more than 1% of the world's electricity, and with plans to place large wind farms in the oceans, that figure will rise. However, the cost of building the wind farms is high, as is the cost of moving electricity from places where the wind is ideal to places where people actually need power.

Ten Ways to Reduce Global Warming

1. Reduce, reuse, recycle

2. Use less heat and air conditioning

3. Replace regular light bulbs with compact fluorescent light (CFL) bulbs

4. Drive less and drive smart

5. Buy energy-efficient products

6. Use less hot water

7. Use the "off" switch

8. Plant a tree

9. Get a home-energy audit from your utility company

10. Encourage others to conserve

Source: www.about.com

Sunlight

Sunlight can be converted into electricity by using specially made solar cells. These have been available for decades, but only recently have prices begun to fall significantly. Because directly generated solar energy is available only while there is daylight, storage is a key factor in how much it can contribute the world's energy usage. Even so, the cost of solar energy looks likely to fall to the point where it becomes a genuine competitor to oil and coal.

Hydroelectric Power

The force of falling water can also be used to produce electricity. Hydroelectric plants use natural waterfalls or artificial dams to provide the rushing water that, like the wind in wind farms, spins the turbines that generate an electric current. They already produce almost 20% of the world's electricity and do not emit any carbon dioxide. But damming rivers can do serious damage to the local environment, and some environmental groups are campaigning for new large hydroelectric projects to be stopped despite their contribution to providing alternative energy.

Biofuels

Scientists are focusing a lot of research on how to use biofuels, which are produced from plants. Ethanol is probably the best known, as it can be produced from such crops as sugar cane, sugar beet and corn and then blended with gasoline to make a fuel that emits less carbon than conventional gasoline. The problem here is that so much energy is used in planting, harvesting and processing these crops that the net saving in energy is less than it appears. By 2008, it was also clear that a sudden rush to produce biofuels meant that farmers were planting fewer crops, pushing up the price of basic foods and causing hardship for poor people around the world.

Nuclear Power

By the middle of the 2000s, people began to look again at nuclear power. No new nuclear power stations have been built in the United States since the end of the 1970s, when a near-catastrophe at the Three

Mile Island plant in Pennsylvania almost killed the nuclear power industry. But using a controlled nuclear reaction to turn water into steam that can drive turbines produces little carbon. Nuclear power stations already produce about 15% of the world's electricity, including more than half the electricity generated in the United States, Japan and France.

Supporters argue that it is cleaner and cheaper than oil or coal. Opponents say that nuclear power is less clean than it looks once the cost of processing the uranium used in the reactors is factored in and that it is not cheaper when the cost of building and maintaining the reactors is taken into account. Moreover, it produces very dangerous radioactive waste that the world still is not completely sure how to deal with. Even so, several countries, including the United States and some in Europe, are looking at increasing their use of nuclear power. And a number of leading environmental groups now back the use of nuclear power, viewing it as a lesser evil than oil or coal and less of a threat than global warming.

Epilogue: Relevance for Today

Climate change will be a significant political and economic issue over the next 10 years. Scientists are likely to produce more evidence that the world is getting warmer largely because of the fuels that humans are burning—though this is still a debate. At the same time, more visible evidence of change—with dramatic shifts in weather patterns in different parts of the world—will probably emerge. That will mean greater pressure from voters for politicians to take tangible steps to tackle global warming. But voters and politicians will both remain wary of measures that risk hurting their economies.

The next protocol after Kyoto, which will be agreed upon by 2012, will set short-term targets for reducing carbon emissions. But its main goal will be to bring the United States, China and India into the U.N. framework for a global approach to climate change.

It is the round of negotiations for the *second* protocol after Kyoto that will show us how well the world has come together to tackle climate change. By that time, around 2020, the impact of climate change on particular regions will be clearer and we will have a much better sense of what problems are to come. Also by then, we are likely to see a new trend of national legislatures taking matters into their own hands and setting legally binding emissions levels that apply to their particular country. The task for the United Nations will then be to coordinate these separate moves so that together they make up an effective international response.

In the meantime, the main task for governments is to encourage us all to use less energy—a task made easier as prices rise due to both the scarcity of, and the increased development costs for, most existing energy sources.

TELL ME MORE ...

4 Books/websites to read

Heat: How We Can Stop the Planet from Burning, George Monbiot (Penguin, 2007)

The Hot Topic: What We Can Do About Global Warming, Gabrielle Walker and Sir David King (Harcourt, 2008)

The Weather Makers: Global Warming and What It Means for Life on Earth, Tim Flannery (Grove, 2007)

www.unfccc.int, U.N. Framework Convention on Climate Change

3 Carbon-producing fuels to learn about

Coal – Coal is a combustible black or brownish-black sedimentary rock composed mostly of carbon and hydrocarbons. It is the most abundant fossil fuel produced in the United States, which actually has 25% of global coal reserves.

Oil – Oil is formed through the heat and pressure of layers of mud and earth on the remains of plants and animals from millions of years ago. It is found in liquid form.

Natural Gas – The main ingredient in natural gas is methane, a gas composed of one carbon atom and four hydrogen atoms. Like coal and oil, natural gas also originates from the remains of plants and animals and requires millions of years to form under heat and pressure.

2 Quotes to note

"If we keep emitting greenhouse gases at current rates, we will see bigger changes this century than we did in the previous century. The amount of warming will depend on choices human beings make."

IPCC, February 2007

"If we [do] not take action to solve this crisis, it could indeed threaten the future of human civilization. That sounds shrill. It sounds hard to accept. I believe it's deadly accurate. But again, we can solve it."

Al Gore, June 2006

1 Myth: Flying accounts for a large percentage of CO_2 emissions ...

While some environmental campaigners have identified air travel as a major contributor to atmospheric carbon, it is actually responsible for only 2 to 3% of CO2 emissions. However, jet exhaust also contains other harmful chemicals, which means the greenhouse impact of jets is actually much greater—2.7x larger, according to the IPCC.

BUSY PERSON'S *intro*
STORY

The world and our lives move at a faster and faster pace. Access to information is greater than ever before—yet there is much we regret not knowing. Too often we have a sense that we *ought* to know more about the world and events around us.

We've noticed that as we develop personally and professionally, our interests grow. We have an emerging desire to know more about the key events, players and issues that shape the world we live in. We become more aware of important issues that we just don't *get*—but at the same time, our lives are busier than ever before.

Whether it's commuting, answering e-mails on your BlackBerry or driving the kids to school, life has kept you from learning about the things that you now sense are significant—and that everyone else is talking about. That's why we've created **BUSY PERSON'S *intro***.

BUSY PERSON'S *intros* are written by a network of respected academics from around the world who are interested in making their field of expertise available to the rest of us. Approachable, respectable, short—they fit perfectly into those 20-minute gaps that busy people like us have in our day.

There's no need to feel overwhelmed anymore—just browse through the title list, pick a few you like and start filling those knowledge gaps today.